Fernwood Forest

Written by Brenda Boreham
Illustrated by Laura Timmermans

Fernwood Forest

Written by Brenda Boreham

Illustrated by Laura Timmermans

ISBN: 978-1-77174-599-4

Visit: www.strongnations.com

Contact: 1-888-278-2202

Cataloging-in-Publication Data

Title: Fernwood forest / written by Brenda Boreham ; illustrated by Laura TImmermans.

Names: Boreham, Brenda, author. | Timmermans, Laura, illustrator. | Strong Nations Publishing, publisher.

Description: Nanaimo, British Columbia : Strong Nations Publishing, 2022. | 32 pages : colour illustrations ; 8.5 x 11 cm. | By Day and By Night Nature Series ; 3

Summary: "Fernwood Forest focuses on a scene typical of a temperate rainforest habitat, showing how living and non-living things are connected in a healthy ecosystem. Readers observe animals going about their daily activities: caring for their young, searching for food, and responding to the changes in light and temperature throughout the day. Backmatter provides facts about the featured animals and plants and suggests ways to be respectful when visiting wilderness areas. Fernwood Forest is part of the By Day and By Night Nature Series. Books in this series focus on a specific habitat and explore the many connections within a healthy ecosystem. Through lyrical stories, strong illustration support, labelled diagrams, and nature notes, readers learn about living and non-living parts of the habitat."-- Provided by publisher.

Identifiers: ISBN 9781771745994

Subjects: LCSH: Rain forests--Juvenile fiction. | LCSH: Rain forest animals--Juvenile fiction.

Classification: LCC PZ7 1 B674 F47 2022 | DDC jC813/.6-dc23

MIX
Paper from responsible sources
FSC www.fsc.org
FSC® C016245

Strong Nations Publishing Inc.

Proudly printed in Canada

The story that unfolds between these
pages was inspired by the generations of creatures
that have shared their natural habitat with me. It takes
place on Sunday, May 13, 2018—my mother's 93rd birthday.

Brenda

To my husband Gregg, with all my love.

Laura

The wood lies peacefully under a blanket of fading stars. In this forever moment, nothing stirs. The only sound is the rustle of a gentle breeze shifting through the treetops.

A soft, grey light begins to filter through the leafy, green canopy. The moment passes. It is dawn.

The early morning light grows brighter and stronger as the sun rises in the eastern sky.

A barred owl perches silently on an overhanging limb of an old maple tree. He grips the branch tightly through a thick layer of moss and scans the ground below. He watches for small creatures that could be hiding in the dried leaves, needles, cones, and other debris that litter the forest floor.

Continuing its upward journey, the sun spreads light and warmth across the treetops.

With their heads raised high and their ears twitching, a pair of black-tailed deer is on the alert. They stop to sniff the early morning air for signs of danger. Cautiously, the deer wander down the forest trail in search of fresh, green shoots and leaves to nibble on.

It is midmorning, and the sun is higher in the sky. Shafts of light filter through the leafy canopy formed by the tallest trees.

A red squirrel dashes across the trail as the sun shines brightly through the web of leaves and branches overhead. He climbs nimbly up the trunk of an old fir tree and sits chattering loudly on one of the lower branches. He scolds two grey squirrels as they pass by in search of food. He is protecting his nest and the nuts and seeds hidden there.

The late morning sun shines brilliantly in a cloudless sky.

With a sudden rush of wings, beaks, and tails, a group of northern flickers lands on a mossy tree stump. The birds cling to the crumbling bark with their sharp claws. Their long, pointed beaks search the deep cracks and crevices for ants, beetles, and snails.

It is midday. The sun is now directly overhead. The light and heat pass through the canopy to the understory below.

High above the woodland trail, a mother robin perches at the edge of her nest in the warm sunlight. Her hungry chicks open their mouths wide as she feeds them insects and worms.

The early afternoon sun begins to shift into the western sky.

With the sun high overhead, small patches of light now filter through the understory to the forest floor. Garter snakes slither quietly across piles of dried-up leaves to gather at the base of an old stump. They bask happily in the warm sunlight.

As the afternoon sun lowers in the sky, its warmth and light are felt throughout the forest below.

Banana slugs forage for food among the layers of decaying leaves. They use their microscopic teeth to chew up mushrooms, seeds, dead leaves, and moss. Slowly, they graze over and around a fallen branch.

The midafternoon sun continues its journey toward the western horizon.

With a twisting, turning pattern of flight, tiger swallowtail butterflies explore a tiny woodland glade. They flutter down to a small mud puddle. As they prepare to drink, they spread their wings to absorb the heat from the sun.

It is late afternoon. The sun slowly sinks toward the level of the treetops at the western edge of the woods.

A family of cottontail rabbits pauses at the edge of the glade. Their ears are upright and alert. Slowly and cautiously, the rabbits hop here and there among the bushes and tall grasses. They stop to nibble on tender new shoots and leaves.

It is early evening, and the light is dimming. Deep shadows are forming, blurring the edges of the trees and other landmarks.

With the shadows lengthening, carpenter ants leave their nest in an old, rotting stump. They are searching for dead insects and nectar to feed the ants in their colony. They will mark a trail from the food back to the nest for the other ants to follow.

The wood fills with the glow of the setting sun. In this forever moment, not a creature stirs. The only sound is the rustle of a gentle breeze shifting through the treetops.

The sun hesitates for a moment and then sinks quietly out of sight. The moment passes. It is dusk.

About The Pacific Northwest Temperate Rainforest

What is the Pacific Northwest temperate rainforest?

The Pacific Northwest temperate rainforest is the largest temperate rainforest in the world. It stretches along the Pacific coast from Alaska to California. In some areas, it receives up to 200 centimetres (80 inches) of rain in one year. The temperate rainforest is made up of many living and non-living things. Western red cedar, Douglas fir, Sitka spruce, western hemlock, and bigleaf maple are common trees in this type of forest ecosystem.

What are living things?

Living things are made of tiny cells that change food into energy. Living things grow and reproduce. They need air, water, light, and food to survive. Plants and animals are living things.

Fern

Salmonberry

Black Bear

Raven

What are non-living things?

Non-living things do not require food and do not grow or reproduce. Water, light, rocks, and air are examples of non-living things in nature.

Water

Light

Rocks

Air

Barred Owl (pages 6, 7) The barred owl is a large bird that ranges in size from 40 to 60 centimetres (16 to 24 inches) from head to tail. This owl is named for the brown and white striped markings on its body. It has a round head; brown eyes; a small, sharp, yellow beak; and sharp talons. The barred owl is a carnivore. It hunts small animals such as rabbits, mice, snakes, squirrels, frogs, and voles.

Black-Tailed Deer (pages 8, 9) The black-tailed deer is a mammal. It is reddish-brown in the summer and brownish-grey during the winter months. It has a black tail and a white belly. It measures up to 1 metre (3 feet) at its shoulder. The males have antlers. The black-tailed deer is an herbivore. It grazes on grass, blackberries, blueberries, salal, lichens, fresh shoots, and other seasonal plants.

Squirrel (pages 10, 11) Squirrels are mammals. The red squirrel is native to British Columbia. It is reddish-brown with white fur on its belly. It measures 18 to 24 centimetres (7 to 9 inches) in length. Red squirrels prefer to live in forested areas. The eastern grey squirrel is greyish-black and has a large fluffy tail. It is much larger than the red squirrel and grows to 24 to 28 centimetres (9 to 11 inches) in length. The eastern grey squirrel was introduced to Vancouver in 1909 and to Vancouver Island in 1966. It lives in forested areas but has adapted well to living in cities. Squirrels are omnivores. Their diet consists mostly of seeds, nuts, fruit, and berries, but sometimes they eat eggs and insects.

Northern Flicker (pages 12, 13) The northern flicker is a bird. It is a member of the woodpecker family. It is brown with black bars on its wings and back. It has a black patch on its breast and black spots on its belly. Its tail is dark on top. Adults are 28 to 36 centimetres (11 to 14 inches) in length. The northern flicker is an omnivore. It uses its long beak to catch flying insects such as butterflies, flies, and moths. It also eats ants, beetles, snails, seeds, nuts, berries, and fruit.

Robin (pages 14, 15) The robin is a bird. It is a member of the thrush family. The robin is greyish-brown with orange on its breast and belly. It has a dark-coloured head and white patches under its tail and on its belly. Adults are 24 to 28 centimetres (9 to 11 inches) in length. The robin is an omnivore. It hunts for whatever is seasonally available in the area. These foods include berries, fruits, worms, insects, snails, and frogs.

Garter Snake (pages 16, 17) The garter snake is a reptile. It is covered in scales. Adult garter snakes are 58 to 76 centimetres (23 to 30 inches) in length. Most garter snake species have stripes that run the entire length of their bodies. The garter snake is a carnivore. It doesn't eat while it hibernates in the winter, but it eats whatever small creatures it can catch during the rest of the year, including slugs, frogs, mice, voles, and earthworms.

Banana Slug (pages 18, 19) The banana slug is a gastropod. It can grow to be 25 centimetres (10 inches) in length. It is sometimes bright yellow but can also be dark green or brown. The banana slug has a thick layer of sticky mucous on its body to protect it from drying out. The banana slug is an omnivore. It eats fruit, berries, mushrooms, leaves, and dead plants and animals.

Western Tiger Swallowtail Butterfly (pages 20, 21) The western tiger swallowtail butterfly is an insect. It is mostly yellow with black markings. It measures 7 to 10 centimetres (3 to 4 inches) from wingtip to wingtip. It has three body parts: head, thorax, and abdomen. The western tiger swallowtail butterfly is an herbivore. It sips the nectar from flowers, such as milkweed, ironweed, and wild cherry. It also sips water from mud puddles to get the salt and minerals it needs.

Eastern Cottontail Rabbit (pages 22, 23) The eastern cottontail rabbit is a mammal. It is grey or reddish-brown, depending on the season. It has long ears; large hind feet; and a short, fluffy, white tail. It grows 36 to 48 centimetres (14 to 19 inches) in length. The eastern cottontail rabbit is an herbivore. It eats whatever plants are available in the area, including fruit, seeds, leaves, bark, and new spring buds and shoots.

Carpenter Ant (pages 24, 25) The carpenter ant is an insect. It varies in colour from black to brown to red. It has three body parts: a heart-shaped head, a thorax, and an abdomen. It grows up to 2.5 centimetres (1 inch) in length. The carpenter ant is an omnivore. It forages for dead insects, honey, and nectar.

Scientists have divided the Pacific Northwest temperate rainforest into three layers: the canopy, the understory, and the forest floor.

Canopy

The canopy is the leafy top layer of the forest. It receives the most sunlight. The tops of the tallest coniferous trees are at this level. Some of these trees grow over 76 metres (250 feet) tall and 4.5 metres (15 feet) across. The canopy provides shade for the plants and animals living below. The shade also helps the soil hold moisture. Many birds live in this layer.

Understory

The understory is the middle layer of the forest. Many smaller trees adapt to living in the shade of the understory.

Forest Floor

The forest floor is the lowest layer of the forest. This layer receives the least amount of sunlight. It is littered with dried leaves, twigs, branches, and fallen trees, which slowly decay and return nutrients to the soil. Small plants, ferns, lichens, and mosses grow here. Most of the forest animals live on the forest floor.

Western Red Cedar

Douglas Fir

The Douglas fir tree can grow 60 metres (200 feet) tall. It is an evergreen tree—its flat needles stay green all year. The young Douglas fir tree has smooth, grey-brown bark. As the tree ages, its bark thickens and develops deep ridges. Its seeds ripen in cones.

Broadleaf Maple

The broadleaf maple can grow 24 metres (80 feet) tall. It is a deciduous tree that sheds its leaves in the fall. The broadleaf maple's brown bark forms narrow ridges. Its pale yellow flowers bloom in the spring. Its seeds, which have two "wings," develop in the fall.

Western Red Cedar

The western red cedar tree can grow as tall as 60 metres (200 feet). It is an evergreen tree—it keeps its yellowish-green leaves all year. It has shreddy, reddish-brown bark. Its seeds form in small cones.

Sword Fern

The sword fern is one of 30 native fern species in Western Canada. It has dark green fronds that grow 1 metre (3 feet) in length. The sword fern reproduces from spores, which develop in double rows on the underside of the fronds. The sword fern is evergreen.

Salmonberry

The salmonberry shrub grows from 1.8 to 2.4 metres (6 to 8 feet) high. Its green leaves grow in groups of three. It is a deciduous bush and loses its leaves in the fall. It has red flowers and soft berries that range in colour from yellow to red. The salmonberry reproduces from seeds.

Vanilla Leaf

The vanilla leaf plant grows one large leaf (divided into three smaller leaflets) at the top of a thin stem. It grows about 30 centimetres (12 inches) high. It has a thin spike covered in tiny, white flowers that appear in the spring. The vanilla leaf reproduces from seeds or by spreading roots.

Welcome to the Pacific Northwest Temperate Rainforest

The plants and animals of the Pacific Northwest temperate rainforest are amazing! They have learned to survive the changes in temperature, daylight hours, and the amount of precipitation that falls throughout the four seasons. Enjoy visiting their homes, and please show respect for all living things.

- The forest is home to many living things. We stay on designated trails to prevent damaging this delicate ecosystem.
- We only pick things up if we are sure they are not living (feathers, nuts, cones, stones, fallen leaves, etc.). We return these things to the forest floor before we go home.
- Plant collection is not permitted in any provincial or national parks. We leave the plants, seeds, and flowers undisturbed.
- We treat all creatures with respect. We observe, but we do not approach them or touch them. We keep our pets on a leash at all times.
- We check the B.C. Parks website for campfire bans or closures before we enter the forest.
- We have campfires only in designated places, and we follow the campground safety rules.
- If we spot a wildfire in British Columbia, we report it immediately: 1-800-663-5555.
- When it is time to leave the forest, we take away everything we brought with us.